MW01632084
TEACHER
RECOMMENDED
Kids
WINTER
ACADEMY
ARGOPREP
5 DAYS A WEEK
2 WEEKS
Mathematics
Social studies
English
Fitness
Science
Mazes
GRADE
2

ArgoPrep is one of the leading providers of supplemental educational products and services. We offer affordable and effective test prep solutions to educators, parents and students. Learning should be fun and easy! To access more resources visit us at www.argoprep.com.

Our goal is to make your life easier, so let us know how we can help you by e-mailing us at: info@argoprep.com.

How to access video explantions?

Download our app: ArgoPrep Video Explanations to access videos on any mobile device or tablet.

You also can access it on our website:

Step 1 - Visit our website at: www.argoprep.com/k8
Step 2 - Click on the **Video Explanations** button located on the top right corner.
Step 3 - Choose the workbook you have and enjoy video explanations.

ISBN: 9781951048600
Published by Argo Brothers.

KIDS WINTER ACADEMY

Kids Winter Academy by ArgoPrep covers material learned in September through December so your child can reinforce the concepts they should have learned in class. We recommend using this particular series during the winter break. This workbook includes two weeks of activities for math, reading, science, and social studies. Best of all, you can access detailed video explanations to all the questions on our website.

Want an amazing offer?

Online premium content includes practice quizzes and drills with video explanations and an automatic grading system.

Chat with us live at www.argoprep.com or email at: info@argoprep.com for this exclusive offer

7 DAY ACCESS
to our online premium content at:
www.argoprep.com

BOOKS BY ARGOPREP

Here are some other test prep workbooks by ArgoPrep you may be interested in. All of our workbooks come equipped with detailed video explanations to make your learning experience a breeze! Visit us at www.argoprep.com

COMMON CORE MATH SERIES

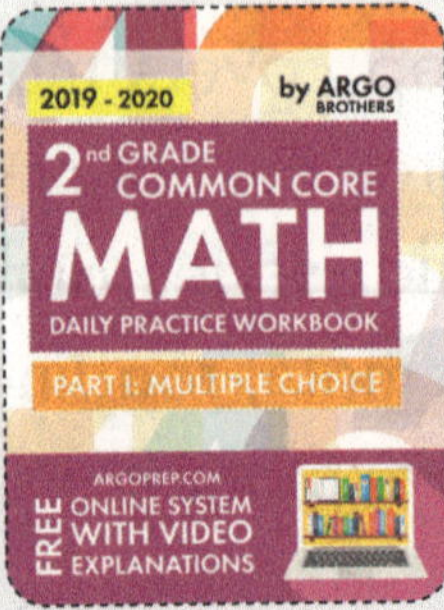

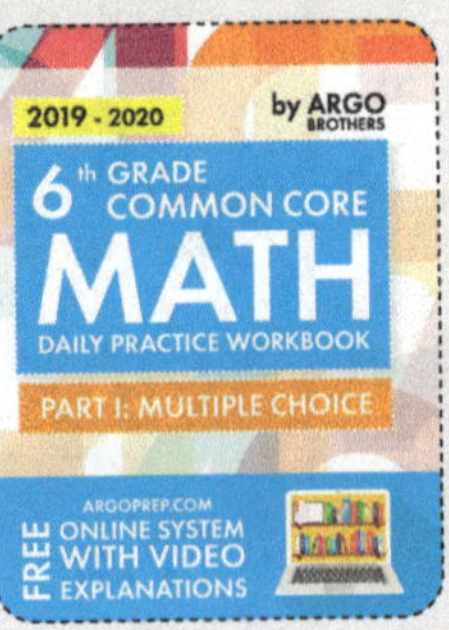

COMMON CORE ELA SERIES

SCIENCE SERIES

Science Daily Practice Workbook by ArgoPrep is an award-winning series created by certified science teachers to help build mastery of foundational science skills. Our workbooks explore science topics in depth with ArgoPrep's 5 E'S to build science mastery: Engaging, Exploring, Explaining, Experimenting, and Elaborating. All of our curriculum is aligned with the latest Next Generation Science Standards.

INTRODUCING MATH!

Introducing Math! by ArgoPrep is an award-winning series created by certified teachers to provide students with high-quality practice problems. Our workbooks include topic overviews with instruction, practice questions, answer explanations along with digital access to video explanations. Practice in confidence - with ArgoPrep!

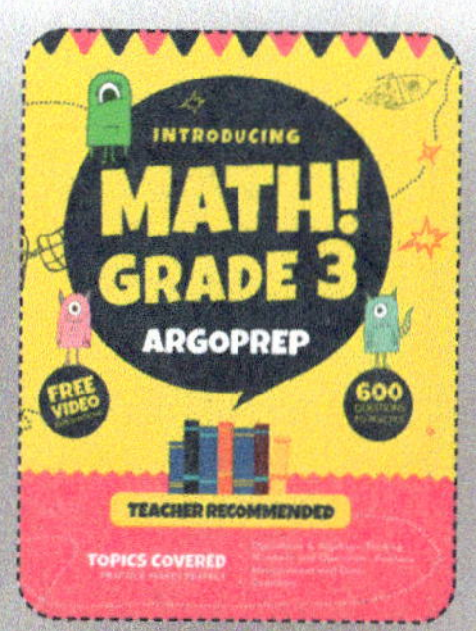

KIDS SUMMER ACADEMY SERIES

ArgoPrep's Kids Summer Academy series helps prevent summer learning loss and gets students ready for their new school year by reinforcing core foundations in math, english and science. Our workbooks also introduce new concepts so students can get a head start and be on top of their game for the new school year!

SUMMER ACTIVITY PLAYGROUND SERIES

Summer Activity Playground is another summer series that is designed to prevent summer learning loss and prepares students for the new school year. Students will be able to practice math, ELA, science, social studies and more! This is a new released series that offers the latest aligned learning standards for each grade.

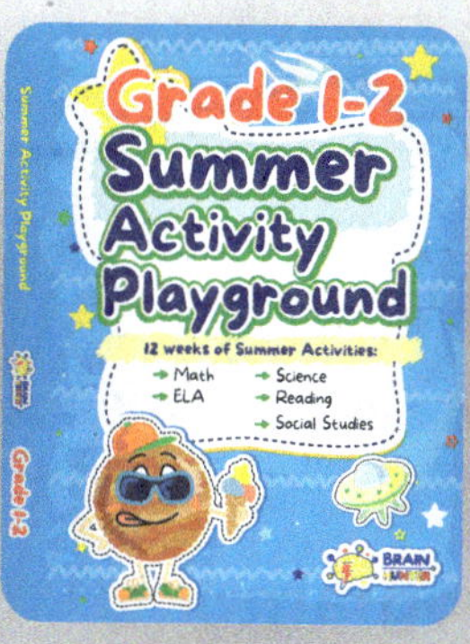

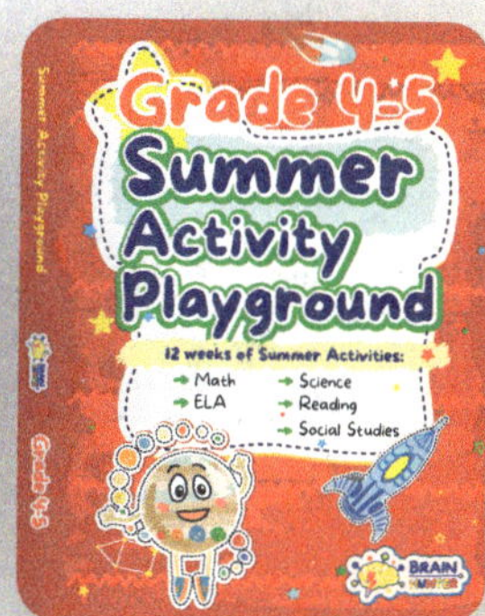

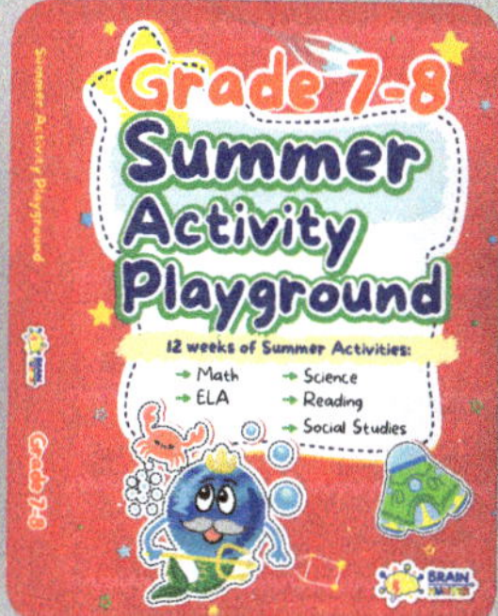

COMMON CORE TEST SERIES

The goal of these workbooks is to provide mock state tests so students can increase confidence and test scores during actual test day.

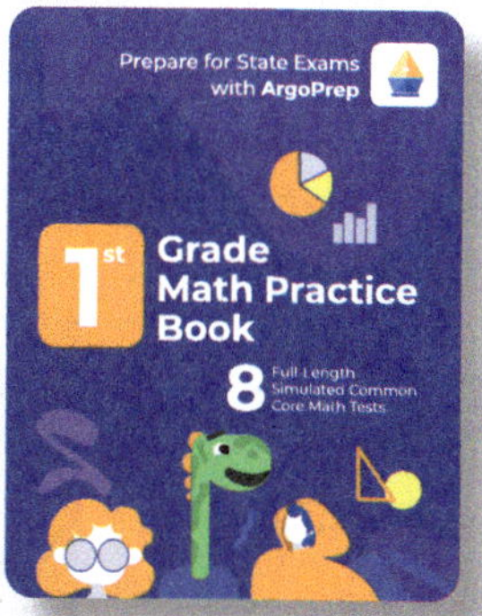

Take a look inside!

Your Child Moves Ahead of the Class With ArgoPrep

K-8 Math and ELA Video Lectures

Your subscription includes ALL grades so you can access video lessons from different grades. We cover and teach every topic your child needs to know for their grade level! All of our video lessons are taught by licensed-teachers and the videos are designed to be engaging!

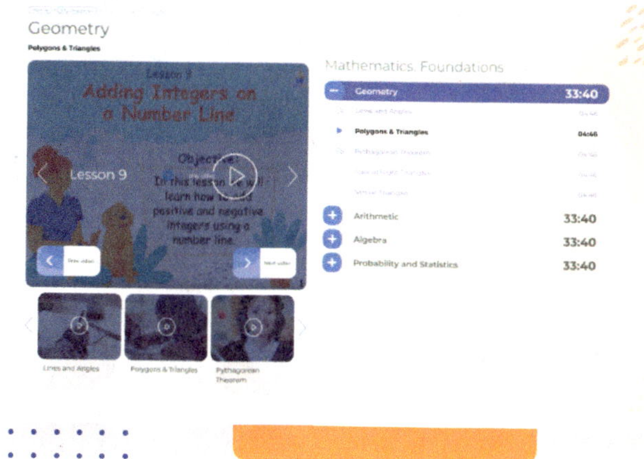

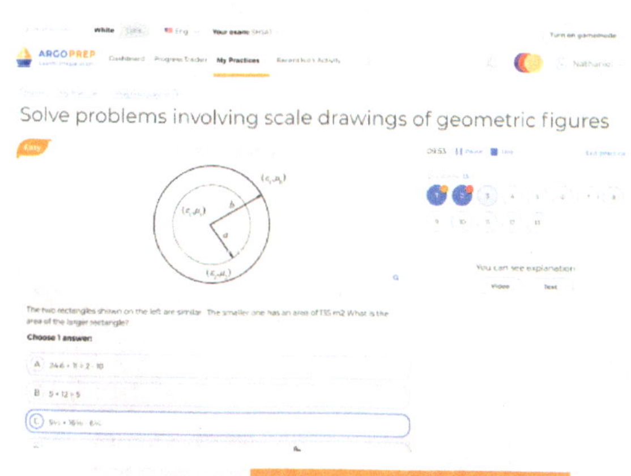

Gap-Proofing Quizzes

Our quizzes gauge student mastery level of any particular topic. If your child struggles, each quiz question has an explanation video to accompany it, so students don't fall through those "learning gaps". If they need more information, they can always review last year's videos and worksheets as well.

Unlimited Printable Worksheets

Print your worksheets from our database of thousands! We are constantly creating new, educator-approved worksheets for grades K-8 in our ever-expanding resource collection! Our worksheets are unique because we include fewer questions and more visually balanced spaces. Our amazing illustrations are 100% kid-approved!

ARGOPREP

Winter Academy

Place Value

Sample Problem: Bubba the Bear munches on 562 berries every morning. Can you break this number down into hundreds, tens, and ones?

.... hundreds

.... tens

.... ones

Answer: 5 hundreds, 6 tens, 2 ones

Explanation: To understand big numbers, take a look at the place value of each digit. From right to left, the order of place value is ones, tens, hundreds. (Hint: Saying the number out loud can be helpful!)

Question 1:

Every afternoon, Bubba loves to take a nap for 245 minutes. Break this number down into hundreds, tens, and ones.

A. 2 hundreds, 4 tens, 5 ones
B. 2 ones, 4 tens, 5 hundreds
C. 4 tens, 2 ones, 5 hundreds
D. 5 tens, 2 hundreds, 4 ones

Question 2:

Bubba slowly climbed to the top of a tree that was 108 feet high. Break this number down into hundreds, tens, and ones.

A. 1 one, 0 tens, 8 hundreds
B. 1 hundred, 0 tens, 8 ones
C. 8 ones, 0 hundreds, 1 ten
D. 1 ten, 0 ones, 8 hundreds

Question 3:

When Bubba roars loudly, his friends can hear him from 610 feet away. Break this number down into hundreds, tens, and ones.

.... hundreds
.... tens
.... ones

Question 4:

Bubba's favorite lunch is a fresh salmon that is at least 850 centimeters long. Write this number out into hundreds, tens, and ones.

....

Reading Comprehension

Directions: Read the passage. Then answer the questions that follow.

Today was the big day! Sari was going horseback riding for the first time. She had been looking forward to this day since the first time she asked her mom to take her riding a few months ago.

As they drove down the long dirt road toward the ranch, Sari felt butterflies in her stomach. Even though she was so excited to ride a horse, she was nervous, too. Horses were so much bigger than her!

The ranch was beautiful! They had horses, ponies, goats, chickens, and pigs. There was a huge saddle barn where all the horses' accessories were kept. Jack, the owner of the ranch, led Sari and her mom into the barn and showed her how to choose a helmet. He explained that they had already assigned each rider to a horse. The horses would be brought out, one by one, for the riders to mount from a wooden platform nearby.

When it was her turn, Sari walked slowly up the steps to the top of the platform. She watched her horse, Annie, be led out of the saddle barn toward her. She looked at her mom with big, round eyes. Could she do it?

Finally, Annie was next to her, and Jack directed Sari on how to mount, or get on, the horse. Sari followed the directions and mounted Annie like a pro. She could hardly believe it! Before long, all the riders had mounted their horses, and they were off, following Jack and his horse along the trail. Sari had never felt so proud of herself!

1. What does the phrase "butterflies in her stomach" mean?

A. Sari saw a butterfly nearby.

B. Sari was feeling nervous.

C. Sari had eaten a butterfly.

2. Why did Sari look at her mom with "big, round eyes" before getting on her horse?

A. She couldn't believe how big the horse was.

B. She was feeling surprised.

C. She was feeling nervous.

3. Who was the main character in the passage?

A. Sari

B. Jack

C. Annie

Classifying Materials by Observable Properties

Directions: Read the passage. Then answer the questions that follow.

"

Just about any material can be classified by its observable properties. Observable properties are the characteristics of different materials that you can observe with your five senses such as color, texture, flexibility, weight, size, shape, and hardness.

For example, if you were classifying a leaf, you might make the following observations about its characteristics:

1. Color: green
2. Texture: smooth
3. Flexibility: bendable
4. Weight: very light
5. Size: small, about the size of a hand

Other materials may have some of the same characteristics even though they are very different objects. Also, sometimes objects that are the same can have some characteristics that are different. For example, leaves can be different colors and shapes.

"

1. Observable properties are observed using the five senses.

   A. True B. False

2. Which of the following statements is true? Circle all correct answers.

   A. All materials have the same characteristics.

   B. Like objects can have different characteristics.

   C. Different objects can have similar characteristics.

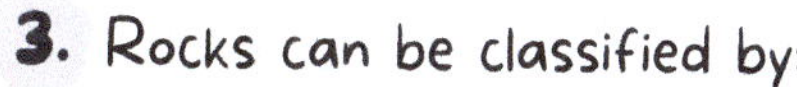

3. Rocks can be classified by:

   A. Weight B. Shape C. Both A and B

4. How might you describe the texture of a rock?

....

Early Settlers

Directions: Read the passage. Then answer the questions that follow.

Following the explorers who came to America, early settlers established the first settlement--a community of people--in Jamestown, Virginia in 1607. Native Americans were already living in America at the time. The Jamestown Settlement was made up of over 200 men who came to America from England.

The settlers built a fort along the water that could be easily protected from enemies. Unfortunately, this site was not ideal. In the summer, the area flooded and became swampy. In the winter, there was no protection from the bitter cold and storms. Most of the settlers did not survive their first winter in Jamestown.

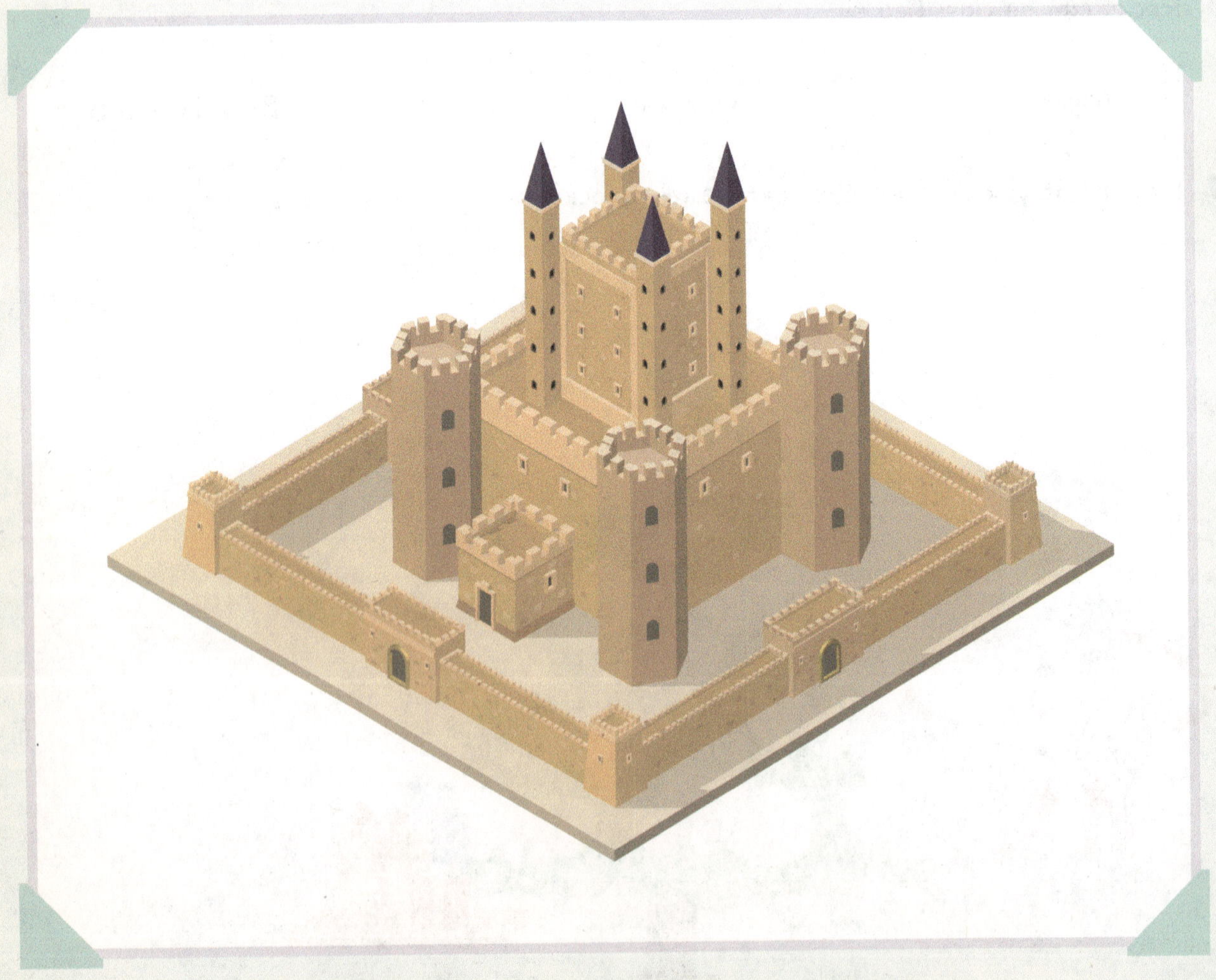

Week 1 Day 1 SOCIAL STUDIES

1. The Jamestown settlers were the first people to come to America.

A. True B. False

2. What does the word "settlement" mean?

A. Explorers C. Community

B. Enemies D. People

3. Why did the settlers build their fort along the water?

A. So they would have water to drink.

B. So the area would become flooded.

C. So they could easily protect it from enemies.

D. Because it was the first place they saw when they arrived.

4. Why did very few settlers survive their first winter in Jamestown?

A. Because the area became flooded.

B. Because they were attacked by enemies.

C. Because there was no protection from the cold and winter storms.

D. Because they had no food to eat.

Skip-Counting with 2s, 5s, 10s, and 100s

Sample Problem: Wanda the Wolf loves to skip-count out loud! While walking through the snowy woods, she skip-counts by 2. She starts at 2 and stops at 20. Write the numbers, Wanda would have said:

Answer: 2, 4, 6, 8, 10, 12, 14, 16, 18, 20

Explanation: Skip-counting is a quick way to count up. In this problem, we're adding 2 each time to our number. You can also skip-count by 5's, 10's, and 100's!

Question 1:

Wanda is counting fish in the stream. She skip-counts by 2. She starts at 6 and stops at 20. What is a number she would not have said?

A. 6
B. 8
C. 16
D. 17

2+2=4

A B C D E

Question 2:

Wanda counts some small flowers in a meadow. They grow in clumps of 5, so she skip-counts by 5. She starts at 5 and stops at 85. What is a number she would not have said?

A. 50
B. 83
C. 25
D. 5

Question 3:

Wanda is watching a bee hive and counting the bees out loud. The bees travel in groups of 10, so she skip-counts by 10. She starts at 80 and stops at 160. What are the numbers Wanda will say out loud?

....

Question 4:

Wanda is running fast across the snow. She skip-counts by 10 as she runs. She starts at 15 and stops at 85. What numbers are missing from her count below?

15, 25, 35, 45, ..., 65, ..., 85

Reading Comprehension

Directions: Read the passage. Then answer the questions that follow.

"

It was a cool evening in early autumn. I took my book out to the porch to read while the kids played with the neighbors next door. As I sat reading, with cars passing by on the street every few minutes, I could hear the kids hosting a talent competition with one another. They were singing and dancing, laughing, and occasionally arguing with one another.

After about an hour, as the sun began to set, I yelled to the kids to come inside.

"But Mom," argued Karis, "we're having so much fun! Just a few more minutes please?"

"OK, fine," I responded, agreeing to give them ten more minutes to play.

"Thanks, Mom," Karis exclaimed, hugging me and running back to her friends next door.

"

1. What is the setting of this story?

A. In a house, during the fall
B. Outside, during the summer
C. At the park, during the fall
D. Outside, during the fall

2. Which of the following is a detail from the passage?

A. Mom was watching TV in the house.
B. The kids were having a talent competition.
C. The kids were not having fun playing together.
D. The weather was rainy and windy.

3. What might be the title of this passage?

A. Mom and Me
B. What is Autumn?
C. Playing with Friends
D. Singing and Dancing

4. The kids in the passage were having fun.

A. True
B. False

Changes Caused by Heating and Cooling

Directions: Read the passage. Then answer the questions that follow.

Most of the time, heating or cooling materials will cause changes within them. Sometimes these changes can be reversed and sometimes they can't. Think about what happens to a glass of water that is put in the freezer for several hours. The water freezes into ice. If the glass is removed from the freezer and left to return to room temperature, the ice will melt back into water.

For some materials, changes from temperature cannot be reversed. Think about what happens when a bag of popcorn is put in the microwave for several minutes. The heat from the microwave causes the corn kernels to expand and pop open. If the popcorn is then exposed to cooler temperatures, the popcorn cannot be changed back into kernels.

1. Some materials can change when exposed to hot or cold temperatures.

   A. True B. False

2. When wood is exposed to fire (heat), it will turn to ash. This change is:

   A. Able to be reversed B. Not able to be reversed

3. What causes corn kernels to pop open?

   A. Cold temperatures B. Hot temperatures

4. When exposed to cold temperatures, water turns to ....

   When exposed to hot temperatures, ice turns back to ....

   A. Water, ice B. Ice, boiling water C. Ice, water

Life in the Past

Directions: Read the passage. Then answer the questions that follow.

"There are many things that are different about life today compared to the past. First, people wore very different clothing in the past, and they sewed many garments (clothing) by hand. Homes also looked different long ago. Many homes did not have indoor bathrooms or kitchens with refrigerators and ovens like we have now. Instead, people cooked food over an open fire.

Like homes, schools also looked different from the school buildings we have now. Most schools had only one room and all children learned together from the same teacher. Children usually walked to school each day, sometimes having to travel long distances. Life looked very different in the past from how it does now!

"

Week 1 Day 2 SOCIAL STUDIES

1. In the past, most clothes were handmade.

   A. True B. False

2. The word "garments" means:

   A. Ovens B. Schools C. Clothing D. Homes

3. Which of the following is NOT a difference between life in the past and now?

   A. Schools had only one room and one teacher.

   B. Homes did not have refrigerators and ovens.

   C. Most clothing was sewn by hand.

   D. Children rode to school in cars not busses.

4. Think of one other aspect of life that was different long ago. In a complete sentence, write it on the lines below.

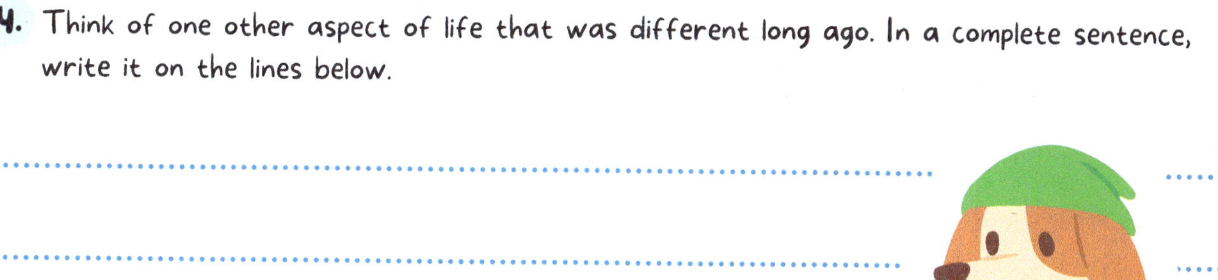

Simple Addition within 20

Sample Problem: Buster the Bunny has a sweet tooth! For breakfast, he eats 6 jelly donuts, 5 chocolate donuts, and 2 powdered donuts. How many donuts did Buster eat?

Answer: 13

Explanation: This is a simple addition problem. You can make a drawing to help you ... or you can practice your mental math!

Question 1:

Buster loves cupcakes. At a friend's birthday party, he gobbles up 7 chocolate cupcakes, 2 vanilla cupcakes, and 8 carrot cupcakes. How many cupcakes did Buster eat?

A. 17
B. 15
C. 10
D. 9

Question 2:

Buster is fond of having cookies for his snack. One afternoon, he eats 4 chocolate chip cookies, 12 peanut butter cookies, and 1 oatmeal cookie. How many cookies did Buster eat?

A. 15
B. 16
C. 13
D. 17

Question 3:

Buster's favorite part of dinner is dessert. One evening he dives into 5 scoops of mint chocolate ice cream, 3 scoops of vanilla ice cream, and 1 scoop of strawberry ice cream. How many scoops did Buster enjoy?

....

Question 4:

At midnight, Buster sneaks several candy bars out of his mother's chocolate stash. He quickly eats 2 milk chocolate bars, 6 dark chocolate bars, and 3 white chocolate bars. How many chocolate bars will his mother be missing in the morning?

....

Reading Comprehension

Directions: Read the passage. Then answer the questions that follow.

I am doing a research report for school on my favorite animal, the sloth. My family took a trip to the zoo so I could see these animals up close. While we were there, we attended a presentation about sloths, and I learned so much about these amazing creatures!

Sloths are a type of mammal that live in Central and South America. Sloths mainly eat the buds, shoots, fruits and leaves of one kind of tree - the Cecropia tree. This tree is also the primary home to sloths. They only leave the tree once per week!

Sloths have long, sharp claws that help them climb trees. Even though they are good at climbing, these cute little animals are actually one of the slowest moving animals in the world. They only move 2 meters per minute! They also sleep a lot - about 10 hours per day.

Learning about sloths at the zoo helped me write a great research report for school!

1. Which is a FACT from the passage?

   A. Sloths move around very quickly.

   B. Sloths eat the leaves and fruits of the tree they live in.

   C. Sloths leave the Cecropia tree once a day.

2. Which is an OPINION from the passage?

   A. Sloths are cute little animals.

   B. Sloths are one of the slowest moving animals in the world.

   C. Sloths are a type of mammal.

3. What was the main idea of this passage?

   A. Everyone loves sloths.

   B. Sloths are interesting animals.

   C. Sloths live at the zoo.

4. Sloths sleep over 10 hours per day.

   A. True B. False

Pollination

Directions: Read the passage. Then answer the questions that follow.

"

Pollination is the process that allows plants to reproduce, or grow new plants. It is very important because without pollination helping plants make new seeds, we would not have fruit to eat or beautiful flowers to look at.

Pollination happens when the pollen from one plant is transferred, or moved, to another plant. Plants rely on bees, butterflies, hummingbirds, and other insects to pollinate. Without these important pollinators, plants would not be able to make seeds and reproduce.

"

1. Which of the following is a true statement from the passage?

A. Bees and butterflies are not an important part of the pollination process.

B. Plants make new seeds and reproduce through the process of pollination.

C. Fruits are delicious to eat.

2. What is the purpose of this passage?

A. To get the reader to like fruit

B. To persuade the reader to help save bees

C. To teach the reader about pollination

3. What does the word "reproduce" mean in the passage?

A. To grow new plants **B.** To move **C.** To pollinate

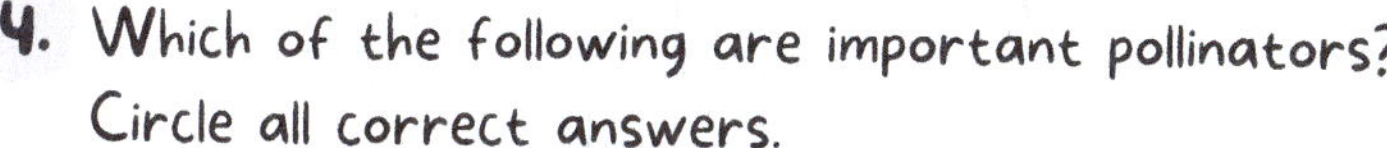

4. Which of the following are important pollinators?
Circle all correct answers.

A. Spiders

B. Bees

C. Hummingbirds

D. Butterflies

E. Squirrels

Equal Rights

Directions: Read the passage. Then answer the questions that follow.

The United States was founded on the belief of equal rights for all of its citizens. Citizens are people with rights and responsibilities in a particular city, state, country, or other community. Rights are freedoms that citizens are entitled to as members of a particular community. As citizens of the United States, we have the right to vote and freedom of speech and religion, among many other things.

As citizens, we also have responsibilities. Responsibilities are the things that citizens are required to do as members of a particular community. These responsibilities include being respectful of our neighbors and taking care of our community by cleaning up trash and not littering. It is important that we always remember our responsibilities as citizens of our communities.

1. A right is:

   A. Something we are entitled to as citizens

   B. Something we are required to do as a member of our community

2. Being kind to our fellow citizens is an example of a:

   A. Right B. Responsibility

3. Our country was founded on the belief of equal rights for all citizens.

   A. True B. False

4. Freedom to own property, or land, is an example of a:

   A. Right B. Responsibility

5. Which of the following is NOT an example of a responsibility?

   A. Cleaning up our trash after a picnic at the park

   B. Being respectful of neighbors by asking to use their swing set

   C. Freedom of speech

   D. Being kind to others in our community

Simple Subtraction within 20

Sample Problem: It's a windy day, and Wanda the Wolf is frustrated. She's setting out all the supplies to make sandwiches, but everything keeps blowing away. She had 12 slices of bread, but 9 blew away. How many slices does she have now?

Answer: 3

Explanation: This is a simple subtraction problem. You can make a drawing to help you ... or you can practice your mental math!

Question 1:

Wanda places 7 slices of turkey on a piece of bread, but then 3 blow away in a gust of wind. How many slices of turkey are left on the piece of bread?

A. 2
B. 3
C. 4
D. 5

Question 2:

Wanda places 15 leaves of lettuce on top of the turkey slices, but then the wind picks up again. Sadly, 3 leaves of lettuce fly off to the right, and 3 leaves fly off to the left. How many leaves of lettuce are left on Wanda's sandwich?

A. 9
B. 12
C. 8
D. 7

Question 3:

Wanda always puts a lot of cheese on her turkey sandwiches. She has 11 slices of cheese to put on top of the lettuce. However, 2 slices drop in the dirt and 1 slice blows away. How many slices of cheese does Wanda have left for her sandwich?

....

Question 4:

Wanda howls in frustration. She was almost ready to eat her sandwich when 7 of her 9 napkins blew out of her lap. How many napkins does Wanda have left to wipe her mouth?

....

Reading Comprehension

Directions: Read the passage. Then answer the questions that follow.

"

There are five steps to making a peanut butter and jelly sandwich. First, you need to gather all the ingredients - bread, peanut butter, jelly, a knife, and a plate. Next, you lay two pieces of bread next to each other on the plate. Then, you use the knife to spread peanut butter on one piece of bread and jelly on the other. After that, you put the two pieces of bread together to create a sandwich. Finally, you clean up after yourself by putting all the ingredients away and the knife in the dishwasher. Now you can sit down and enjoy the delicious peanut butter and jelly sandwich you worked hard to make! Yum!

"

1. According to the passage, there are 3 steps to making a peanut butter and jelly sandwich.

   A. True   B. False

2. What do you do after you clean up after yourself?

   A. Gather the ingredients
   B. Put the pieces of bread together
   C. Enjoy the sandwich
   D. Spread the jelly on the bread

3. What is the purpose of this passage?

   A. To entertain the reader
   B. To persuade the reader to make a peanut butter and jelly sandwich
   C. To teach the reader how to do something

4. Which of the following is an example of a transition word from the passage? Circle all correct answers.

   A. First   B. Finally   C. Six   D. Then

5. An appropriate title for this passage could be, "How to Make a Peanut Butter and Jelly Sandwich."

   A. True
   B. False

Diversity of Life in Different Habitats

Directions: Read the passage. Then answer the questions that follow.

There are many different types of habitats, or homes, that animals live in, such as woodlands, wetlands, oceans, and rainforests. Every habitat is home to a variety of animals, plants, food, water, trees, and shelter. Different types of plants and animals live in different habitats depending on their needs.

For instance, woodlands are full of animals such as squirrels, chipmunks, deer, birds, and small insects, as well as plants such as trees, flowers and grasses. Most woodland animals live in or near trees and eat nuts, fruits, seeds, and grasses. These things are found in abundance in woodland areas, making this a perfect habitat for those animals. Animals such as dolphins, zebras, and polar bears have different needs. They are not found in woodland areas because they would not be able to survive.

1. A habitat must be able to meet an animal's .... for it to be able to survive in that environment.

A. Food B. Water C. Needs D. Space

2. All habitats are home to the same types of plants and animals.

A. True B. False

3. Which types of animals would probably be found in a wetlands area?

A. Raccoons and chipmunks
C. Polar bears and fish
B. Fish and frogs
D. Sloths and monkeys

4. Which of the following can be found in a habitat? Circle all correct answers.

A. Plants B. Animals C. Food D. Cars

Good Citizenship

Directions: Read the passage. Then answer the questions that follow.

We are all citizens of many different communities. A citizen is someone with rights and responsibilities in a particular city, state, country, or other community. Citizens have rights and freedoms that they are entitled to as a citizen of a community. They also have responsibilities-- things they should do as a member of a community.

People who are good citizens typically have character traits that make them admirable. They take their responsibilities as a citizen seriously by treating other members of the community with kindness and respect. They also take care of their community by ensuring it is clean, safe, and welcoming to all. Good citizens make a community a better place to live, work, and learn!

SOCIAL STUDIES

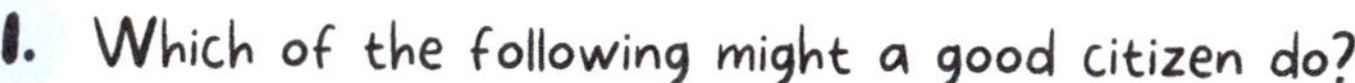

1. Which of the following might a good citizen do?

   A. Throw trash on the ground

   B. Be mean to a neighbor

   C. Help a neighbor carry their groceries in the house

   D. Make others feel unwelcome

2. You can only be a member of one community.

   A. True   B. False

3. Which communities are you a citizen of? Circle all that apply.

   A. Neighborhood   C. State   E. School

   B. City   D. Country   F. Religious organization

4. Which of the following would NOT be a character trait of a good citizen?

   A. Rude   B. Thoughtful   C. Friendly   D. Kind

Money

Sample Problem: It's a snowy day, and Buster the Bunny, Wanda the Wolf, and Bubba the Bear decide to go to the movies. Bubba has brought 3 quarters, 1 dime, 1 nickel, and 1 penny to spend on popcorn. How much money does he have?

Answer: 91 cents

Explanation: Knowing what different coins are worth is an important life skill. One quarter is worth 25 cents. One dime is worth 10 cents. One nickel is worth 5 cents. One penny is worth 1 cent. So for this problem, the addition would look like this: 75 + 10 + 5 + 1 = 91

Question 1:

Wanda is hoping to buy a fizzy orange drink. She checks her pocket and finds 1 quarter, 2 dimes, and 3 nickels. How much money does she have to spend?

A. 80 cents
B. 75 cents
C. 50 cents
D. 60 cents

Question 2:

Buster has headed straight to the candy bars in the movie theater. He has 6 dimes, 3 nickels, and 5 pennies to spend. How much money is that?

A. 60 cents
B. 70 cents
C. 80 cents
D. 75 cents

Question 3:

Bubba is ready to buy his movie ticket. It will cost 2 quarters, 1 dime, 1 nickel, and 2 pennies. What is the price of the movie ticket?

....

Question 4:

Buster spent almost all of his money on candy bars. Wanda gives him 1 quarter, 2 dimes, and 3 pennies to help him pay for his movie ticket. (Buster promises to share his candy bars.) How much money did Wanda give him?

....

Reading Comprehension

Directions: Read the passage. Then answer the questions that follow.

In 1872, Yellowstone National Park became the first national park to be formed in the United States. It was named after the Yellowstone River which flows through the park. Although almost the entire park is located in Wyoming, there are small sections found in both Idaho and Montana.

Yellowstone National Park is most famous for the Old Faithful Geyser which is found in the park. A geyser is a hot spring that shoots water and steam up into the air. About half of the world's geysers are located in Yellowstone National Park.

Yellowstone National Park is also home to grizzly bears, wolves, bison, and elk. People come from all over the world to visit Yellowstone National Park each year.

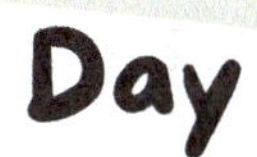

1. A geyser is:

   A. A waterfall

   B. A hot spring that shoots out water and steam

   C. A river

2. Yellowstone National Park spans .... different states.

   A. 1 B. 2 C. 3

3. About $\frac{1}{3}$ of the world's geysers are found in Yellowstone National Park.

   A. True B. False

4. Yellowstone National Park is a popular place for people to visit.

   A. True B. False

5. Old Faithful is:

   A. A famous hot spring B. A famous river C. A famous geyser

Erosion

Directions: Read the passage. Then answer the questions that follow.

Erosion is the process of the land changing shape or being worn away through natural forces such as wind, water, and rain. Erosion usually happens over hundreds, thousands, and even millions of years. Erosion has created many beautiful formations such as mountains, cliffs, canyons, and rock formations. The Grand Canyon was created by erosion.

Erosion can be harmful in many ways. For instance, erosion can cause the top layer of soil to wash away on farmland. This layer is the most nutrient-rich, so when it is washed away, poor quality soil is left behind. This can affect the crops grown by farmers, and ultimately, the food we eat.

1. Erosion happens:

A. Over a short period of time
B. Over a long period of time

2. What causes erosion?

A. Wind
B. Water
C. Rain
D. All of the above

3. Erosion is always harmful.

A. True
B. False

4. Erosion impacts many people.

A. True
B. False

5. Erosion is the process of:

A. Land changing shape through natural forces
B. Mountains being formed
C. Farmers losing crops
D. Weather changing

The Pledge of Allegiance

Directions: Read the passage. Then answer the questions that follow.

The Pledge of Allegiance was written in 1892 by Francis Bellamy, a minister from New York. It was written for a celebration in honor of the 400th anniversary of Columbus Day. The Pledge of Allegiance was first recited at that celebration. Today, these words are still recited in many schools across the United States every morning:

"I pledge allegiance to the flag of the United States of America, and to the republic for which it stands, one nation under God, indivisible, with liberty and justice for all."

Week 1 Day 5 SOCIAL STUDIES

1. The Pledge of Allegiance was written for a celebration honoring:

   A. The President
   B. The United States of America
   C. Francis Bellamy
   D. Columbus Day

2. The Pledge of Allegiance was written by:

   A. A minister
   B. The President
   C. A teacher
   D. Columbus

3. The Pledge of Allegiance is still recited in schools today?

   A. True
   B. False

4. What does it mean to "pledge allegiance"?

   A. To like something
   B. To be part of something
   C. To want something
   D. To promise to be loyal

Merry
Christmas

Merry
X-mas

ARGOPREP

Grade 2

WEEK 2

Winter Academy

Addition with 3-Digit Numbers

Sample Problem: Marvin the Moose loves to stroll through fresh snow. He makes 216 tracks in the morning and 115 tracks in the afternoon. How many tracks did he make total?

Answer: 331 tracks

Explanation: You'll need your skills in addition to solve this type of problem. Remember to line up your numbers according to place value and then add.

$$\begin{array}{r} 216 \\ +\,115 \\ \hline \end{array}$$

Question 1:

One of Marvin's favorite treats is an apple right off the tree. One afternoon, he ate 105 red apples and 176 green apples. How many apples did Marvin eat?

A. 281
B. 218
C. 210
D. 181

MATH

Question 2:

Sometimes Marvin's back gets itchy. One morning he scratched his back 411 times up against a pine tree and then 330 more times against an oak tree. How many times did Marvin scratch his back?

A. 71
B. 74
C. 714
D. 741

Question 3:

It's no wonder Marvin's back is itchy. He counted 505 fleas on his upper back and 303 fleas on his lower back. How many total fleas did Martin count?

....

Question 4:

For once, Marvin has decided to take a bath with soap. He scrubs for 295 seconds, rinses off, and scrubs for another 705 seconds. How many seconds does Marvin scrub up?

....

Reading Comprehension

Directions: Read the passage. Then answer the questions that follow.

Hiking is a great way to spend time in nature and get your body moving. One of the best things about hiking is that you really don't need a lot of expensive or special gear. Most people get by with a backpack, water bottle, and good sneakers.

Finding places to hike just about anywhere is fairly easy as well. Checking out nearby state parks is a great place to start. Every state has a wide variety of state parks, and most have marked hiking trails that range in difficulty from very easy to very rugged. Be sure to choose the appropriate level of trail for your ability. It's also important to keep in mind that some trails may be several miles long and take many hours to hike.

Regardless of the type and length of trail you hike, you will most likely be rewarded with beautiful scenery, magnificent animal and insect sightings, and the sense of calm that comes from being in nature.

1. Most hikers need expensive gear.

   A. True
   B. False

2. Which of the following statements is true? Circle all correct answers.

   A. Hiking is a great form of exercise.
   B. Finding places to hike is difficult.
   C. Being in nature is calming for many people.

3. What was the main purpose of the passage?

   A. To entertain the reader
   B. To inform the reader
   C. To persuade the reader

4. Where does the author suggest people hike?

   A. City parks
   B. State parks
   C. National parks

Water on Earth

Directions: Read the passage. Then answer the questions that follow.

Almost $\frac{3}{4}$ of the Earth is covered by water, and 97% of this water is found within the world's five oceans. These are the Atlantic, Pacific, Indian, Arctic, and Southern (Antarctic) Oceans. Although all living things on Earth need water to survive, ocean water is salt water and is unsuitable, meaning not good, for drinking or growing crops.

The remaining water on Earth can be found in freshwater lakes, ponds, and rivers. This water can be used for drinking and farming but is in a much smaller supply. Some freshwater is even frozen solid in the form of glaciers-- slow moving masses of ice.

Arctic Ocean
Atlantic Ocean
Pacific Ocean
Pacific Ocean
Indian Ocean
Southern (Antarctic) Ocean

1. Some of the water on Earth is solid and some is liquid.

A. True

B. False

2. What does the word "unsuitable" mean?

A. Good for

B. Not good for

C. Full of

D. Salty

3. The Earth is made up of .... water.

A. 75%

B. 97%

C. 3%

4. Most of the water on Earth can be found in:

A. Lakes

B. Rivers

C. Glaciers

D. Oceans

The World on a Map

Directions: Read the passage. Then answer the questions that follow.

A world map is a flat drawing that helps us locate places around the world. It shows all seven continents and five oceans, as well as the Equator, the North and South Poles, and many other things.

Maps include a compass rose and a legend to help us make sense of the information included on the map. A compass rose helps us to know which direction - north, east, south or west - is which on the map. A legend, sometimes called a map key, helps us to identify the different symbols used on a map. For instance, a blue line may be the symbol used to show rivers on a map.

Maps are very important tools because they help us to locate places all over the world!

Week 2 Day 1 SOCIAL STUDIES

1. A map is different from a globe.

A. True
B. False

2. Which is the same as a legend on a map?

A. The Equator
B. A compass rose
C. A map key
D. Symbols

3.  .... shows us directions on a map.

A. The Equator
B. A compass rose
C. A map key
D. Symbols

4. We use the .... to help us identify different symbols on a map.

A. compass rose
B. legend
C. world map
D. equator

Subtraction with 3-Digit and 2-Digit Numbers

Sample Problem: Rosie the Raccoon is very sneaky. She spots a pile of 150 marshmallows next to a campfire and sneaks off with 23 of the marshmallows. How many were left in the pile?

Answer: 127 marshmallows

Explanation: You'll need your skills in subtraction to solve this type of problem. Remember to line up your numbers according to place value and then subtract.

$$\begin{array}{r} 150 \\ -\ 23 \\ \hline \end{array}$$

Question 1:

Rosie goes back to the campfire to look for other treats. She spots a stack of 334 graham crackers and sneaks off with 90 of them. How many graham crackers did she leave behind?

A. 244
B. 144
C. 225
D. 110

Question 2:

There was a pile of 167 sticks next to the campfire. Rosie takes 83 sticks off the pile. How many were left in the pile?

A. 84 sticks
B. 80 sticks
C. 87 sticks
D. 100 sticks

Question 3:

Still looking for goodies, Rosie finds 498 chocolate bars inside a backpack. She manages to grab 232 of the chocolate bars before she hears campers coming. How many chocolate bars are still in the backpack?

....

Question 4:

Rosie quickly hides just before 562 campers come back to the campfire. When they discover things missing, 200 of the campers decide to stay at the campfire while the rest go in search of the thief. How many campers are going to be looking for Rosie?

....

Reading Comprehension

Directions: Read the passage. Then answer the questions that follow.

Noah loves playing basketball. He started playing when he was just five years old and has played every year since. This year, his team is the Rockets. Most of the other players on the team are new to the sport. He knows he is the best player on the team, but he doesn't like to brag, so he tries to help teach the other kids so they can improve.

Noah also really likes his coach, Coach Damien. Coach Damien knows a lot about basketball and teaches Noah many new drills he has never done before. Noah enjoys going to practice each week. Games are his favorite, though. Noah loves the feeling he gets when he shoots a basket during a game and the whole crowd cheers.

Noah works hard during practices because he really wants to play basketball in high school and college. He even dreams of becoming a professional basketball player, but he knows he has to be diligent about practicing and improving. He will work hard, though, to reach his goals.

1. Who is the main character in the story?

   A. Noah B. Coach Damien C. A basketball player

2. What does the word "diligent" probably mean?

   A. Funny B. Hard working C. Smart

3. According to the story, which is Noah's favorite?

   A. Basketball practice B. Basketball drills C. Basketball games

4. Which of the following best describes Noah?

   A. Hard working and kind

   B. Athletic

   C. Smart and compassionate

Volcanoes

Directions: Read the passage. Then answer the questions that follow.

A volcano is an opening in the Earth's surface that can sometimes become active. When a volcano becomes active, gas, ash, and magma escape in the form of eruptions. Some volcanic eruptions have sent ash into the air as far as 17 miles away. This can be extremely dangerous for buildings, homes, people, and animals nearby.

75% of the volcanoes on Earth can be found in an area of the United States near the Pacific Ocean called the Pacific Ring of Fire. Mount St. Helens is one such volcano located in this area. This volcano had a major eruption in 1980. While most volcanoes are found on land, there are even some volcanoes on the ocean floor!

Week  2 Day  2  SCIENCE 

1. Volcanoes can be very dangerous.

   A. True   B. False

2. Most volcanoes can be found in:

   A. North America
   B. Central America
   C. South America
   D. Canada

3. Which of the following erupts out of a volcano?

   A. Gas
   B. Ash
   C. Magma
   D. All of the above

4. Which of the following is a FACT from the passage?

   A. The Pacific Ring of Fire is located in Asia.
   B. Mount St. Helens has never erupted.

   C. Volcanoes can be found on the ocean floor.

   D. All volcanoes are found on top of mountains.

Map Symbols

Directions: Read the passage. Then answer the questions that follow.

All maps have symbols. Map symbols are pictures, drawings, colors, or patterns that represent something real. They are used to help us identify places or things on a map. For instance, a small symbol like this may represent a house on a map, while a symbol like this may represent a park.

A key to the symbols used on a map can be found in the legend, or map key. This legend is usually a small box in the corner of the map that shows each symbol and what it represents. When looking for places on a map, it is important to use the map key to help you learn what each symbol represents.

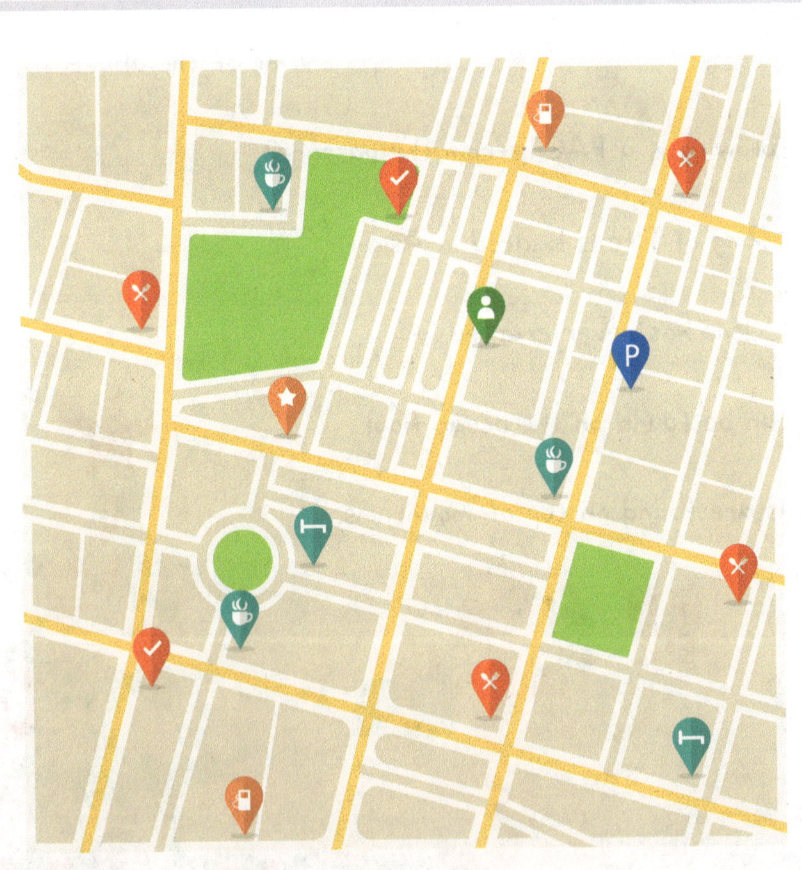

1. A  helps you to know what each symbol stands for.

   A. Map
   B. Symbol
   C. Picture
   D. Map key

2. Map symbols can be:

   A. Pictures
   B. Drawings
   C. Patterns
   D. All of the above

3. What might this symbol represent on a map?

   A. Store
   B. Library
   C. Park
   D. Office building

4. Draw a symbol below that could be used to represent an airport on a map.

Ordering Numbers Least to Greatest

Sample Problem: Eddie the Eagle has carefully counted his feathers. He has **55** gray feathers, **19** white feathers, **22** speckled feathers, and **43** black feathers. Order these numbers from least to greatest.

Answer: 19, 22, 43, 55

Explanation: Thinking about where numbers are on a number line will help you order any group of numbers from least to greatest.

Question 1:

Eddie has finished his super-size nest in time for winter. He's used **29** large branches, **56** medium branches, **13** small branches, and **97** tiny branches. Order these numbers from least to greatest.

A. 97, 56, 29, 13
B. 13, 29, 56, 97
C. 13, 56, 97, 29
D. 29, 56, 97, 13

MATH

Question 2:

Eddie can spot lots of creatures in the icy river below him. He sees 15 trout, 16 salmon, 8 frogs, and 35 tadpoles. Order these numbers from least to greatest.

A. 8, 15, 16, 35
B. 35, 16, 15, 8
C. 8, 16, 15, 35
D. 8, 35, 15, 16

Question 3:

Eddie flies over a snowy meadow and sees more wildlife. He sees 65 mice, 5 moles, 17 bunnies, and 1 gopher. Order these numbers from least to greatest.

....

Question 4:

When he's feeling bored, Eddie screeches. He screeched 29 times on Monday, 45 times on Tuesday, 17 times on Wednesday, and 11 times on Thursday. Order these numbers from least to greatest.

....

Reading Comprehension

Directions: Read the passage. Then answer the questions that follow.

Lila's favorite thing to do with her mom is make homemade pizza. The whole family loves it so much that they usually have it every Friday night!

The first thing Lila and her mom do is make the pizza dough. It only takes 4 ingredients-- flour, yeast, water, and salt! Once the dough is ready, Lila uses a rolling pin to flatten the dough into a large circle. While she does that, her mom prepares the other ingredients. When the dough is rolled out into a big enough circle, Lila pours pizza sauce on top and spreads it out with a spoon. After that, she and her mom sprinkle mozzarella cheese on top and layer the pizza with pepperoni, their favorite topping.

After all the ingredients are on the pizza, it goes into the oven for 20 minutes. Lila and her mom set the table while the pizza is baking. Once it's done, they let it cool and then use a pizza cutter to cut the pizza into 8 pieces-- 2 pieces for each person in their family. Finally, it's time to sit down at the table and enjoy their hard work!

1. Which of the following is a true statement from the passage?

   A. Lila rolls the dough out into a large rectangle.

   B. Lila's favorite pizza topping is pepperoni.

   C. They cut the pizza into 10 pieces.

2. What is the purpose of this passage?

   A. To get the reader to like pizza

   B. To persuade the reader to make pizza

   C. To teach the reader how to make pizza

3. When making pizza, which step comes right before layering on the pepperoni?

   A. Sprinkling the pizza with mozzarella cheese

   B. Rolling out the dough

   C. Spreading on the pizza sauce

4. Fact or opinion? <u>Pepperoni is the best pizza topping.</u>

   A. Fact

   B. Opinion

Cloud Types

Directions: Read the passage. Then answer the questions that follow.

Clouds are made up of tiny droplets or frozen crystals of water. Most clouds are formed when warm air rises into the atmosphere-- the air above Earth-- and cools down into tiny droplets of water or ice. As more droplets are formed, they come together to form a cloud.

There are four main types of clouds: cirrus, cumulus, stratus, and nimbus. Cirrus clouds are thin, wispy clouds found high up in the sky. Cumulus clouds are puffy clouds scattered throughout the sky. Stratus clouds look like a large blanket covering the sky. They are usually a good indicator of rain or snow. Nimbus clouds are seen during a thunderstorm. They are dark and already have rain or snow falling from them.

1. Which type of cloud can be found high up in the sky?

A. Cirrus
B. Cumulus
C. Stratus
D. Nimbus

2. What is the purpose of this passage?

A. To get the reader to like clouds

B. To persuade the reader to look at clouds

C. To teach the reader about the different types of clouds

3. Which type of cloud already has water falling from it in the form of rain or snow?

A. Cirrus
B. Cumulus
C. Stratus
D. Nimbus

4. What can we conclude about the weather when the sky is full of stratus clouds?

A. It will be a nice day.

B. It will be a rainy or snowy day.

C. It will be a windy day.

D. It will be a clear day.

Compasses

Directions: Read the passage. Then answer the questions that follow.

A compass is a tool used for finding direction. It is small and round with a magnetic needle mounted on a pin. The needle always points north which is helpful for people trying to navigate, or find their way through, a new area. A magnetic compass is similar to, but not exactly the same as, a compass rose found on a map.

Many different types of people find compasses useful. Hikers can use them while hiking through the woods so they do not get lost. Sailors use them to navigate the seas. Miners use compasses to get around underground. Without compasses, many people would find it difficult to get around.

1. A compass and a compass rose are the same thing.

A. True B. False

2. Who might find a compass useful?

A. Hikers
B. Miners
C. Sailors
D. All of the above

3. On a compass, the needle always points to the:

A. North B. East C. South D. West

4. Without compasses, many people could get ....

A. Help B. Hungry C. Lost D. Home

More Than/Less Than/Equal Symbols

Sample Problem: Marvin the Moose and Rosie the Raccoon are on a bowling team together and have their own bowling balls. Marvin's ball weighs 16 pounds and Rosie's ball weighs 9 pounds. Which of these comparisons is true?

16 > 9
16 < 9
16 = 9

Answer: 16 > 9

Explanation: Explanation: The symbol > means "is more than." The symbol < means "is less than." The symbol = means "is equal to." Now it's just a matter of finding the symbol that makes the comparison true. (Hint: Imagine the < or > symbol to be a hungry alligator's mouth that's wide open. Now remember the alligator always wants to "eat" the bigger number!)

Question 1:

Marvin scored 119 in the first bowling game. Rosie scored 211. Which of these comparisons is true?

A. 119 > 211
B. 119 < 211
C. 119 = 211
D. None of the above

Question 2:

Marvin scored 115 points and then 10 points was added to his total score. Rosie scored a 129. Which comparison is correct?

A. $(115 + 10) > 129$
B. $(115 + 10) < 129$
C. $(115 + 10) = 200$
D. None of the above

Question 3:

Marvin scored 198 in the second bowling game. Rosie scored a 200. Write the comparison of their two scores.

....

Question 4:

Marvin scored 185 in the final bowling game. Rosie scored 193 but had 7 points taken away from her score. Write the comparison of their two scores.

....

Reading Comprehension

Directions: Read the passage. Then answer the questions that follow.

Narwhal and Whale lived in the same part of the deep, blue ocean. Whale was jealous of Narwhal because she had a beautiful horn that made her special.

"I wish I had a horn like yours," Whale told Narwhal one day.

"I wish I could give you my horn. I don't like it!" Narwhal replied.

"You don't like it? But why?" Whale asked, feeling astonished.

"Shark and his friends always make fun of my horn, and I feel embarrassed," said Narwhal.

At that very moment, Shark and several of his friends swam up and began laughing and making fun of Narwhal. Narwhal turned bright pink and looked very sad. Whale felt so sorry for Narwhal.

"Shark, why are you making fun of Narwhal?" asked Whale.

"Because she looks so silly with that horn," replied Shark, laughing meanly.

"She does NOT look silly. She looks beautiful and unique. I think you're just jealous of her and that's why you make fun of her," Whale said.

"Jealous? I'm n-not jealous," Shark replied, but he sounded unsure as he and his friends swam away.

Once they were gone, Whale went over to comfort Narwhal.

"I'm so sorry, Narwhal. Now I understand why you don't like your horn. I didn't know you got made fun of for it. I want you to know, though, that I love it and think it makes you unique!"

"Thank you," responded Narwhal, as they swam off to play.

1. This passage is a fable. A fable teaches a moral, or lesson, through the story.

   What is the moral of this story? Circle all correct answers.

   A. Always be kind to others.
   B. Sharks and narwhals can't be friends.
   C. Be thankful for what you have.
   D. Don't be nice to sharks.

2. Why do you think Shark makes fun of Narwhal?

   A. He wants to be her friend.
   B. He wishes he had a horn.
   C. He wants to make Whale angry.
   D. He doesn't like Narwhal.

3. What is the meaning of the word "unique?"

   A. Ugly
   B. Beautiful
   C. Different
   D. Same

4. How does Narwhal feel at the end of the story?

   A. Sad
   B. Angry
   C. Jealous
   D. Happy

5. Narwhal and Whale are friends at the end of the story.

   A. True
   B. False

Classifying Living Things by Physical Features

Directions: Read the passage. Then answer the questions that follow.

Living things such as plants and animals can be classified, or grouped together, by their physical features. Physical features are characteristics that can be observed with the five senses. Examples of physical features include an animal's body covering, body parts, and method of movement.

For instance, a bird and a butterfly could be grouped together as living things with wings. However, they would not be grouped together if you were observing their body coverings because birds have feathers but butterflies do not.

1. Classified means:

A. To observe

B. To group together

C. To look the same

D. To look at

2. Physical features must be observable with the five senses.

A. True

B. False

3. Which types of animals would probably be classified together based on type of movement?

A. Fish and giraffe

B. Frog and bird

C. Hummingbird and skunk

D. Whale and shark

4. Which types of animals would probably be classified together based on body covering?

A. Polar bear and chipmunk

B. Frog and cat

C. Dolphin and snake

D. Fish and polar bear

Producers and Consumers

Directions: Read the passage. Then answer the questions that follow.

"

People need various goods and services to survive. Goods are items that are tangible such as food, clothing, and tools. Services are things other people with certain skills do for us. Common services include car repair, medical help, and haircuts. We usually pay for goods and services with currency (money).

The people who purchase goods and services for their own use are called consumers. Producers are people or companies who make or supply goods to consumers. People can be both consumers and producers, depending on the good or service.

"

1. .... are intangible.

A. Goods B. Services

2. Currency is a synonym for:

A. Goods B. Services C. Money

3. A person can be both a consumer and a producer.

A. True B. False

4. We pay people with certain skills to provide a ....

A. Good B. Service

5. Olivia goes to the store to buy apples, bread, eggs, and milk. She is an example of a ....

A. Producer B. Consumer

Topic: Addition/Subtraction Problems with Lengths

Sample Problem: Marvin the Moose, Rosie the Raccoon, and Eddie the Eagle take turns sledding down a snowy hill on their sleds. Marvin's sled is 6 feet longer than Rosie's sled. If Rosie's sled is 3 feet, what's the total length of Marvin's sled?

Answer: 9 feet

Explanation: In these problems, you'll need to decide whether to add or subtract. If you're looking for the total of two numbers, then you'll need to use addition. If you're looking for the difference between two numbers, then you'll need to use subtraction. We know Marvin's sled is longer, so we need to add 6 feet to 3 feet. That's 9 feet - just enough room for a moose!

Question 1:

Rosie is really fast on her sled and slid a distance of 75 feet. Eddie is a little slower and slid a distance of 63 feet. What is the difference between the two distances?

A. 8 feet
B. 138 feet
C. 12 feet
D. 10 feet

Question 2:

Rosie owns two very fast sleds. One is 85 inches long and the other is 36 inches long. What's the difference in length between the two sleds?

A. 121 inches
B. 94 inches
C. 20 inches
D. 49 inches

Question 3:

Eddie knows he can fly a lot faster than he can sled, but he sticks with it. For his first time sledding down the hill, he went a distance of 45 feet. On his second time, he went 10 feet further. How far did Eddie sled on his second try?

....

Question 4:

Marvin, Rosie, and Eddie sled down the snowy hill all at the same time. Marvin slides a distance of 33 feet, Rosie slides a distance of 40 feet, and Eddie slides a distance of 27 feet. If they added up all of their distances, how far did they slide all together?

....

Reading Comprehension

Directions: Read the passage. Then answer the questions that follow.

Mars is a planet. It is the fourth planet from the sun and the second smallest planet in the solar system. Mars is known as the Red Planet because of its color.

Because Mars is very similar to Earth, scientists have studied the Red Planet for many years. They want to find out if there was ever life on Mars.

Mars is home to the biggest mountain in the solar system known as Olympus Mons. It is three times bigger than Mount Everest, the biggest mountain on Earth. Scientists have also found evidence of water on Mars.

1. Olympus Mons is:

   A. The biggest mountain on Earth

   B. The biggest mountain on Mars

   C. The biggest mountain in the solar system

2. Mars got its nickname from its:

   A. Color B. Size C. Shape

3. Mars is smaller than Earth.

   A. True B. False

4. There is no water on Mars.

   A. True B. False

5. What is the main topic of this passage?

   A. Olympus Mons

   B. Earth

   C. Mars

   D. Mount Everest

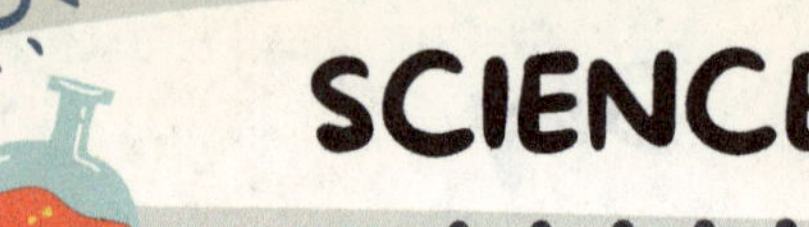

Animal Adaptations

Directions: Read the passage. Then answer the questions that follow.

An **animal adaptation** is a behavior or physical trait that improves an animal's chance of survival in its natural environment. Adaptations do not happen immediately; they occur over time. Examples of animal adaptations include the long necks of giraffes (which allow them to reach leaves high up in trees to eat); the thick layers of fat and dense fur of polar bears (which protect them from the frigid arctic temperatures); and the ability of chameleons to camouflage, or blend in with their surroundings (which keeps them safe from predators). Without these adaptations, none of these animals would be able to survive for very long in their natural environments.

1. An adaptation is:

   A. A change in behavior

   B. A change in physical trait

   C. Both a and b

2. Animal adaptations occur:

   A. Quickly

   B. Immediately

   C. Over a long period of time

   D. Never

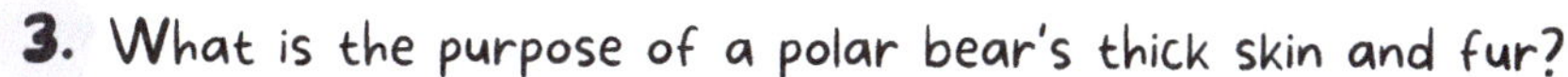

3. What is the purpose of a polar bear's thick skin and fur?

   A. To keep it warm

   B. To protect it from predators

   C. To help it find food

4. Without adaptations, animals would not be able to survive very long in their natural environments.

   A. True

   B. False

5. What is the purpose of a bird's sharp claws?

   A. To help it find a home

   B. To help protect it from predators

   C. To help it capture and eat food

   D. To keep it warm

Scarcity

Directions: Read the passage. Then answer the questions that follow.

People use money to exchange, or pay, for goods and services. What happens when goods --necessary items--are not available or are limited? Scarcity means there is a shortage of a certain good or service. For instance, if a particular city or state is expecting a winter storm, certain items such as bread, milk, and eggs become scarce, or difficult to find at the store.

There are several different situations that can result in scarcity. Sometimes goods can become scarce when too many people want a certain item or when an item cannot be made quickly enough to satisfy demand. Services may be scarce because there are not enough people who want to provide the service.

1. People usually exchange money for goods and services.

A. True B. False

2. Scarcity means an item is:

A. Abundantly available

B. Limited in supply

C. Completely unavailable

3. Services can also be scarce at times.

A. True B. False

4. Goods can be scarce because .... people want a certain item.

A. Not enough B. Just enough C. Too many

5. Services can become scarce when not enough .... are able to provide the service.

A. Goods B. People C. Scarcity

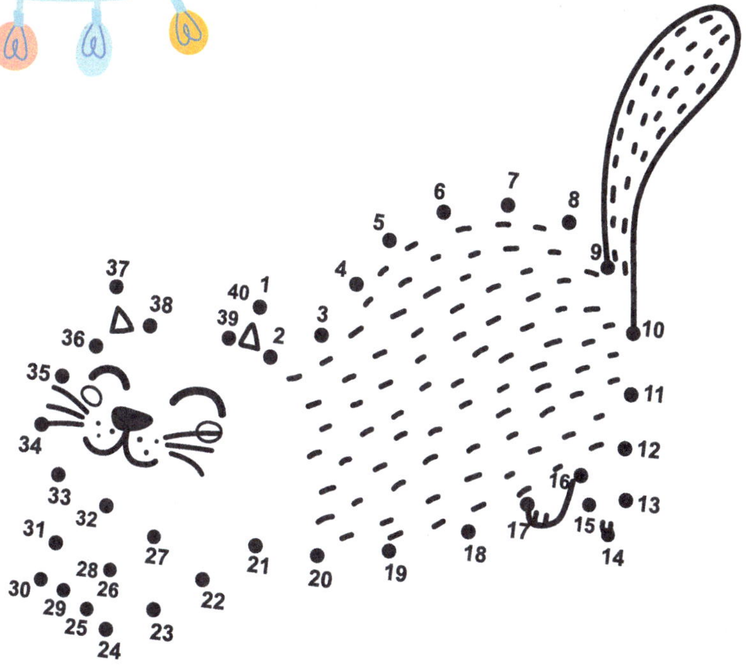

1 2 3 4 5 6 7 8 9 10 11 12 13 14 15 16 17 18 19 20 21 22 23 24 25 26 27 28 29 30 31 32 33 34 35 36 37 38 39 40

FITNESS STUDIES

exercises complex one

1 - **Sit-ups:** 3 times

2 - **Lunges:** 2 times to each leg.
Note: Use your body weight or books as weight to do leg lunges.

3 - **Plank:** 6 sec.

4 - **Run:** 50m
Note: Run 25 meters to one side and 25 meters back to the starting position.

"Please be aware of your environment and be safe at all times. If you cannot do an exercise, just try your best."

exercises complex two

1 - **High Plank:** 6 sec.

2 - **Chair:** 10 sec.
Note: sit on an imaginary chair while keeping your back straight.

3 - **Waist Hooping:** 10 times. Note: if you do not have a hoop, pretend you have an imaginary hoop and rotate your hips 10 times.

4 - **Sit-ups:** 10 times

FITNESS STUDIES

exercises complex three

2 - **Bend Down:** 10 sec.

3 - **Chair:** 10 sec.

1 - **Down Dog:** 10 sec.

5 - **Shavasana:** as long as you can.
Note: think of happy moments and relax your mind.

4 - **Child Pose:** 20 sec.

"Please be aware of your environment and be safe at all times.
If you cannot do an exercise, just try your best."

exercises complex four

2 - **Lunges:** 3 times to each leg.
Note: Use your body weight or books as weight to do leg lunges.

1 - **Bend forward:** 10 times.
Note: try to touch your feet. Make sure to keep your back straight and if needed you can bend your knees.

3 - **Plank:** 6 sec.

4 - **Sit-ups:** 10 times

Grade 2
ANSWER KEY
Winter Academy
ARGOPREP

MATH

Question 1: A

Question 2: B

Question 3: 6 hundreds, 1 ten, 0 ones

Question 4: 8 hundreds, 5 tens, 0 ones

ENGLISH

1. B
2. C
3. A

SCIENCE

1. A
2. B, C
3. C
4. Answers will vary. (sample: smooth, jagged, bumpy)

SOCIAL STUDIES

1. B
2. C
3. C
4. C

MATH

Question 1: D
Question 2: B
Question 3: 80, 90, 100, 110, 120, 130, 140, 150, 160
Question 4: 55 and 75

ENGLISH

1. D
2. B
3. C
4. A

SCIENCE

1. A
2. B
3. B
4. C

SOCIAL STUDIES

1. A
2. C
3. D
4. Answers will vary. (sample - People used horses and carriages to get around long ago because they did not have cars.)

Week 1 Day 3 ANSWER KEY

MATH

Question 1: A
Question 2: D
Question 3: 9 scoops
Question 4: 11 chocolate bars

ENGLISH

1. B
2. A
3. B
4. A

SCIENCE

1. B
2. C
3. A
4. B, C, D

SOCIAL STUDIES

1. A
2. B
3. A
4. A
5. C

Week 1 Day 4 ANSWER KEY

MATH

Question 1: C
Question 2: A
Question 3: 8 slices
Question 4: 2 napkins

ENGLISH

1. B
2. C
3. C
4. A, B, D
5. A

SCIENCE

1. C
2. B
3. B
4. A, B, C

SOCIAL STUDIES

1. C
2. B
3. Answers will vary.
4. A

Week 1 Day 5 ANSWER KEY

MATH

Question 1: D
Question 2: C
Question 3: 67 cents
Question 4: 48 cents

ENGLISH

1. B
2. C
3. B
4. A
5. C

SCIENCE

1. B
2. D
3. B
4. A
5. A

SOCIAL STUDIES

1. D
2. A
3. A
4. D

Week 2 Day 1 ANSWER KEY

MATH

Question 1: A
Question 2: D
Question 3: 808 fleas
Question 4: 1,000 seconds

ENGLISH

1. B
2. A, C
3. B
4. B

SCIENCE

1. A
2. B
3. A
4. D

SOCIAL STUDIES

1. A
2. C
3. B
4. B

Week 2 Day 2 ANSWER KEY

MATH

Question 1: A
Question 2: A
Question 3: 266 chocolate bars
Question 4: 362 campers

ENGLISH

1. A
2. B
3. C
4. A

SCIENCE

1. A
2. A
3. D
4. C

SOCIAL STUDIES

1. D
2. D
3. B
4. Answers will vary. (Sample - an airplane)

MATH

Question 1: B
Question 2: A
Question 3: 1, 5, 17, 65
Question 4: 11, 17, 29, 45

ENGLISH

1. B
2. C
3. A
4. B

SCIENCE

1. A
2. C
3. D
4. B

SOCIAL STUDIES

1. B
2. D
3. A
4. C

Week 2 Day 4 ANSWER KEY

MATH

Question 1: B

Question 2: B

Question 3: $198 < 200$

Question 4: $185 < (193 - 7)$

ENGLISH

1. A, C
2. B
3. C
4. D
5. A

SCIENCE

1. B
2. A
3. D
4. A

SOCIAL STUDIES

1. B
2. C
3. A
4. B
5. B

MATH

Question 1: C
Question 2: D
Question 3: 55 feet
Question 4: 100 feet

ENGLISH

1. C
2. A
3. A
4. B
5. C

SCIENCE

1. C
2. C
3. A
4. A
5. C

SOCIAL STUDIES

1. A
2. B
3. A
4. C
5. B